BEGINNERS GUIDE TO PINEAPPLE FOR PROFIT

Understanding Growth Phases, Pruning Strategies, and Harnessing the Power of Nature in Your Garden

Lucille Sarron

Table Of Contents

CHAPTER 1 6

Pineapple 6

An Overview Of Pineapple Cultivation's History 6

The Value Of Growing Pineapples 7

Financial Possibilities For Novices 8

CHAPTER 2 12

Understanding Pineapple Varieties...................12

Overview Of Pineapple Varieties:..................12

Selecting The Appropriate Type For Your Area: 14

Characteristics And Yield Of Different Varieties: 15

CHAPTER 3 18

Essential Instruments And Equipment 18

1. Essential Equipment For Small-Scale Farming: 18

2. Specialized Equipment For Commercial Plantations:..19

3. Cost-Effective Alternatives For Beginners:20

CHAPTER 4 22

Choosing An Appropriate Site For Growing Pineapples 22

   1. Pineapple Climate And Soil Requirements: 22

   2. Watering And Sunlight Considerations:.....23

   3. Affecting Site Selection Factors: 23

CHAPTER 5 26

Methods For Preparing Soil And Planting..........26

Testing And Amendments For Soils: 26

Planting Crowns Vs. Pineapple Suckers: 27

Methods For Planting And Spacing:................28

CHAPTER 6 32

Proper Nutrition And Fertilization For Growing Pineapple 32

   1. Understanding Pineapple Nutrient Needs: 32

   2. Choosing The Right Fertilizers: 33

   3. Organic Vs. Synthetic Fertilizers: 34

   4. Application Advice: 35

CHAPTER 7 38

Techniques For Irrigation 38

Watering Quantity And Frequency: 38

Systems Of Drip Irrigation: 39

Rainwater Collection For Eco-Friendly Agriculture: 41

CHAPTER 8 44

Controlling Insects And Diseases In Pineapple Production 44

Typical Pests That Damage Pineapples: 44

The Use Of Integrated Pest Management (IPM) And Preventative Measures: 46

Methods For Identifying And Controlling Diseases:...48

CHAPTER 9 50

Techniques For Controlling Weeds In Pineapple Growing...50

Methods Of Manual Weed Control: 50

Methods Of Chemical Weed Control: 51

Mulching Methods For Weed Control:...........52

Weed Management That Is Integrated (IWM): 53

CHAPTER 10 56

Pruning And Thinning Techniques In Pineapple Farming...56

Importance Of Pruning In Pineapple Farming: 56

Fruit Thinning To Enhance Quality: 58

Timing And Methods Of Pruning: 59

CHAPTER 11 62

Harvesting And Handling After Harvest 62

Pineapple Maturity Indicators: 62

Methods Of Harvesting To Get The Highest
Yield: 63

Tips For Handling And Storing After Harvest: 65

Summary..67

Summary Of The Main Ideas: 67

Obstacles And Remedies In The Pineapple
Industry:..69

Motivation For Upcoming Pineapple Growers:
....71

THE END 74

CHAPTER 1

Pineapple

The tropical fruit pineapple (Ananas comosus) is renowned for its tart and sweet flavor. It is frequently grown in tropical and subtropical parts of the world and is a member of the Bromeliaceae family. Growing pineapple for personal or commercial use may be a fulfilling venture. We'll go over the brief history of pineapple farming, its significance, and its potential for financial gain in this conversation, especially for newcomers.

An Overview Of Pineapple Cultivation's History

The history of pineapple agriculture is extensive and goes back many centuries. Pineapples originated in South America, namely in the area that extends from southern Brazil to Paraguay and northern

Argentina. Native Americans dispersed pineapples across the continent. Pineapples were discovered by European explorers, such as Christopher Columbus, who carried them back to Europe.

After that, the fruit spread to other tropical areas including Asia and Africa, and finally the Caribbean. With the creation of sizable pineapple farms, Hawaii emerged as a major participant in the 20th century's pineapple agriculture. Nowadays, pineapple is farmed around the world, with the Philippines, Thailand, Indonesia, and Costa Rica being the major producers.

The Value Of Growing Pineapples

1. Pineapples are a fruit that is high in vitamins, minerals, and antioxidants, making them a nutritious food. In addition to manganese and vitamin C, it has the anti-inflammatory enzyme bromelain.

2. Pineapples are versatile and can be used in a wide range of culinary preparations, such as being eaten raw or added to juices, jams, and desserts. Because of its adaptability, the food industry favors it.

3. Health Benefits: Pineapple's high nutritional content is linked to a number of health advantages, such as strengthened immunity, better digestion, and possibly even anti-inflammatory properties.

4. Agro-Ecological Benefits: By incorporating pineapple plants into agroforestry systems, you can achieve ecological benefits like soil preservation, erosion prevention, and increased biodiversity.

Financial Possibilities For Novices

1. Minimal Initial Investment: Growing pineapples doesn't require a lot of infrastructure, and starting costs can be quite low, particularly for backyard or small-scale farming.

2. Quick Returns on Investment: After planting, pineapple plants typically begin to bear fruit 18 to 24 months later. Beginners seeking a comparatively quick return on their investment may find this quick turnaround appealing.

3. Market Demand: Pineapple products have a steady demand worldwide, which presents openings for new competitors. In addition to being eaten raw, pineapples are processed to make a variety of goods, which increases their market potential.

4. Adaptability: While it does best in tropical regions, pineapples can grow in a variety of soil types and environments. It can be grown in a variety of climates due to its adaptability.

5. Diversification: For small-scale farmers, pineapple growing can be part of a diversified farming method, offering numerous cash sources and decreasing risks associated with depending on a single crop.

In conclusion, cultivating pineapple offers a fascinating journey anchored in history, with a fruit that possesses nutritional, gastronomic, and economic value. For newcomers, the potential for a successful endeavor is clear, considering the relatively low entrance hurdles and the varied market prospects accessible for this tropical delicacy.

CHAPTER 2

Understanding Pineapple Varieties

Overview Of Pineapple Varieties:

Pineapple (Ananas comosus) is a tropical fruit that belongs to the Bromeliaceae family. There are various types of pineapples, each with its distinct traits, tastes, and adaptability for varied growing situations. The most regularly produced pineapple types include:

1. Smooth Cayenne (Queen Victoria):

• Recognized for its cylindrical form and spiky, golden-yellow exterior.

• Delivers a sweet and tangy taste with minimal acidity.

• Widely grown for canning because to its enormous size and good yields.

2. Queen Victoria (Queen):

• Smaller in size compared to Smooth Cayenne, having a sweeter flavor.

• Suggested for preparing pineapple juice and fresh consumption.

• Suitable for small-scale commercial operations and home gardens.

3. MD-2 (Honey Gold, Golden Diamond):

• Well-known for its tiny crown, cylindrical form, and brilliant golden hue.

• Has a sweet taste due to its high sugar content and low acidity.

• Because of its wonderful taste and appearance, it is frequently selected for fresh fruit markets and export.

4. Ethiopia (Eleuthera):

• Distinguished by its reddish-orange exterior, cylindrical shape, and compact size.

• Has a distinct flavor that is sweet and slightly acidic.

• Grown for fresh consumption and local markets in certain tropical regions.

Selecting The Appropriate Type For Your Area:

It's important to take the intended use, soil conditions, and climate into account when choosing a pineapple variety. Considerable elements include:

1. Weather:

• While some types may withstand subtropical settings, others are more suited for tropical regions.

For instance, MD-2 grows well in warmer areas, although Queen Victoria can withstand a wider variety of temperatures.

2. Type of Soil:

Pineapples favor sandy, well-draining soils that range in pH from slightly acidic to neutral.

• Understanding your soil type can influence the choice of a variety that adapts well to specific soil conditions.

3. Planned Use:

• Sweeter cultivars, like MD-2, might be chosen if growing for fresh food.

• For processing or canning, types with large yields and robust features, such as Smooth Cayenne, are commonly preferred.

Characteristics And Yield Of Different Varieties:

1. Smooth Cayenne:

• Characteristics: Large size, low acidity, good yield.

• Yield: Suitable for commercial cultivation due to its prolific fruit production.

2. Queen Victoria:

• Characteristics: Sweet taste, smaller size.

• Yield: Ideal for home gardens and smaller-scale farming.

3. MD-2:

• Characteristics: Golden color, low acidity, high sugar content.

• Yield: Attractive look and flavor make it appropriate for fresh markets and export.

4. Pernambuco:

• Characteristics: Compact size, unusual scent, sweet and slightly acidic flavor.

• Yield: In certain tropical places, locally grown for fresh consumption.

In conclusion, successful pineapple cultivation requires an understanding of the various qualities of pineapple varieties. A pineapple crop that thrives and produces the best quality and yields may be achieved by selecting the appropriate variety depending on the environment, soil, and intended purpose.

CHAPTER 3

Essential Instruments And Equipment

A rewarding endeavor, growing pineapples requires the right tools and equipment for successful cultivation. This article provides a thorough overview of the many types of equipment and tools required for pineapple farming, such as entry-level inexpensive options, specialist equipment for large plantations, and fundamental tools for small-scale farming.

1. Essential Equipment For Small-Scale Farming:

a. Hand Tools: - Garden Spade and Fork: Used for soil preparation and planting. - Pruning shears: necessary for cutting off unwanted shoots and trimming leaves. - Gloves: Guard hands from potentially irritating substances and the rough leaves. - Hoe: Assists in tilling the ground surrounding pineapple plants and removing weeds.

b. Watering Equipment: - Watering Can or Hose: Ensure a consistent and regulated water supply.

C. Using the pH and moisture meters in the soil testing kit, you may monitor and modify the soil's parameters to promote healthy development.

d. Pots or Seed Trays: - For Seedlings: Use to germinate pineapple seeds or nurture miniature plants before transplanting.

2. Specialized Equipment For Commercial Plantations:

a. Drip Irrigation System: - Efficient water usage: Ensures stable moisture levels without water loss.

b. Mechanical Planters: - Saves time and labor: Facilitates speedier planting on a bigger scale.

c. Fertilizer Applicators: - Precision and efficiency: Ensures optimum distribution of fertilizers.

d. Mulching Equipment: - Mulch Spreaders: Speeds up the mulching process for weed control and moisture retention.

e. Harvesting Tools: - Pineapple Slicer or Harvest Knife: Facilitates efficient and cautious harvesting.

3. Cost-Effective Alternatives For Beginners:

a. Handheld Sprayer: - For Small Areas: An cheap solution for spraying pesticides or fertilizers.

b. DIY Mulching: - Use Organic Materials: Collect leaves, straw, or grass clippings for mulching.

c. Compost made at home: - Compost kitchen waste: By producing nutrient-rich compost at home, you may cut back on the need for pricey fertilizers.

d. Simple Weeding equipment: Use simple weeding equipment to manually manage weeds instead of spending a lot of money on pesticides.

e. Build Your Own Rainwater gathering System: - Gather Rainwater: Lower your water bills by installing a basic rainwater gathering system.

Recall that your budget and the size of your pineapple farm will determine the equipment you choose. To guarantee the lifespan and effective operation of your instruments, routine cleaning and maintenance are vital. The success of your pineapple farming endeavor will depend on your investment in the appropriate tools and equipment, regardless of experience level or intention to build a commercial plantation.

CHAPTER 4

Choosing An Appropriate Site For Growing Pineapples

1. Pineapple Climate And Soil Requirements:

• Climate: Tropical and subtropical regions are ideal for pineapple growth. They like temperatures between 65°F and 95°F (18°C and 35°C), however they are vulnerable to frost. Make sure the weather is consistently warm and free of frost.

• Soil: Sandy-loam soils with good drainage are preferred by pineapples. The optimal pH range for soil is 4. 5 to 6.5. Although they can tolerate a variety of soil types, they dislike wet circumstances. The addition of organic matter to the soil increases drainage and fertility.

2. Watering And Sunlight Considerations:

• Sunshine: Because they are sun-loving plants, pineapples need six to eight hours a day of direct sunshine. Make sure there is enough sunshine in the selected area for healthy growth and fruit development.

• Watering: Because of their shallow root systems, pineapples are prone to waterlogging. Water regularly and moderately, especially during dry spells. But keep in mind that overwatering might result in root rot. It's important to maintain the soil continuously damp but not soggy.

3. Affecting Site Selection Factors:

• Topography: To avoid water stagnation, select a location with high parts or slopes that drain effectively. This aids in preventing root problems brought on by an abundance of moisture.

• Wind Exposure: Pineapples, especially when young, are vulnerable to wind damage. To protect the plants from heavy gusts, choose a spot with some wind protection, such as among natural barriers or windbreaks.

• Closeness to the Equator: Because the equator provides constant sunshine for the whole year, pineapples thrive there. With the right environment, they may be grown at lower latitudes, though.

• Accessibility: Take into account how easily accessible the area is for regular upkeep duties like weeding, fertilizing, and harvesting. Having easy access makes taking care of the plants more convenient.

• Soil Testing: Before planting, test the soil for pH and nutrient levels to see if any amendments are needed to suit the unique nutritional requirements of pineapples.

• History of Pests and illnesses: Look into the past of the local illnesses and pests. The requirement for insecticides and fungicides can be decreased by choosing a location with a low prevalence of illnesses and pests.

When choosing a location, keep these things in mind to make sure the ideal conditions are met for raising fruitful and robust pineapple plants. To guarantee the success of your pineapple farming, don't forget to often check on the situation and make necessary modifications.

CHAPTER 5

Methods For Preparing Soil And Planting

Yes, let's explore the idea of planting methods and soil preparation for pineapple cultivation. This entails a number of crucial steps, such as evaluating the soil and adding additives, deciding whether to plant pineapple crowns or suckers, and figuring out the best planting techniques and spacing.

Testing And Amendments For Soils:

1. Testing of Soils:

• Soil testing must be done prior to pineapple planting. This will reveal information on the pH, nutrient content, and makeup of the soil. Pineapples grow best in soil that ranges in pH from 5. 5 to 7, which is slightly acidic to neutral.

2. Modifications to the Soil:

• Amendments might be required based on the findings of the soil test. To enhance soil fertility and structure, common amendments include organic matter (compost or well-rotted manure).

• Sand or perlite can help improve drainage if the soil is heavy. Pineapples prefer well-draining soil.

Planting Crowns Vs. Pineapple Suckers:

1. Suckers for pineapples:

• The shoots known as suckers grow at the base of the fully grown pineapple plant. To grow new pineapple plants, they can be carefully removed and replanted.

• Because suckers typically mature faster than crowns, they are frequently chosen for propagation.

2. Crowns of pineapples:

• The mature pineapple's leafy tips are called the crowns. To grow new plants, they can be chopped off and planted.

• Compared to suckers, crowns might take longer to establish and bear fruit.

Methods For Planting And Spacing:

1. Positioning:

It's important to give pineapples enough room to develop and to allow for ventilation. It is often advised to space plants 24 to 30 inches apart in rows and 36 to 48 inches apart.

• Wider spacing promotes disease resistance and let's light penetrate all sections of the plant.

2. Planting Techniques:

Raised Beds: Using raised beds for pineapple planting enhances drainage and lowers the possibility of root rot. The height of the raised beds should be between 8 and 12 inches.

• Mulching: Adding a layer of mulch around the pineapple plants aids in temperature regulation, weed suppression, and soil moisture retention.

3. Depth of Planting:

• Plant crowns or suckers of pineapples so that the root is slightly below the soil's surface. To avoid rotting, don't plant too deeply.

4. Watering

• Pineapples need regular moisture, particularly in the early months. But since they cannot withstand being wet, it is essential to have soil that drains well and to water them sparingly.

5. The process of fertilization

• Use a balanced fertilizer on pineapple plants on a regular basis, especially during the growing season. Based on the requirements of the plant and any inadequacies found during soil testing, modify the fertilizer.

You may start a profitable and healthy pineapple plantation by taking care of the soil, selecting the appropriate planting materials, and using the right spacing and planting procedures. A plentiful crop will eventually result from the successful growth of pineapple plants, which will require regular monitoring and care.

CHAPTER 6

Proper Nutrition And Fertilization For Growing Pineapple

Growing pineapple successfully includes giving correct nutrition and fertilizer to ensure the plants acquire the needed nutrients for maximum development and fruit output. Let's look into the principles of correct nutrition and fertilization, analyzing pineapple nutrient demands, choosing the suitable fertilizers, and the comparison between organic and synthetic fertilizers.

1. Understanding Pineapple Nutrient Needs:

Pineapple plants require a balanced amount of nutrients for proper growth. The essential nutrients needed for pineapples are nitrogen (N), phosphorus (P), potassium (K), magnesium (Mg), sulfur (S), calcium (Ca), and trace elements including iron

(Fe), manganese (Mn), zinc (Zn), and copper (Cu). These nutrients serve vital roles in numerous physiological processes like as photosynthesis, root growth, and fruit production.

2. Choosing The Right Fertilizers:

Pineapples benefit from fertilizers that give a proper ratio of the required elements. It's crucial to pick a fertilizer with a composition appropriate for fruiting plants. Look for a fertilizer with a greater potassium concentration, as this encourages blooming and fruit growth. Balanced N-P-K (Nitrogen-Phosphorus-Potassium) ratios, such as 10-10-20 or 14-14-14, are generally advised for pineapple production.

It's preferable to use slow-release fertilizers to maintain a continuous supply of nutrients over an extended period. This helps avoid over-fertilization, which can lead to nutritional imbalances or environmental contamination.

3. Organic Vs. Synthetic Fertilizers:

• Natural Fertilizers:

• Compost, manure, bone meal, and other naturally occurring materials based on plants or animals are the sources of organic fertilizers.

• As they decompose, they release nutrients gradually, offering a more consistent and mild nutritional supply.

• Organic fertilizers improve the general health of the soil by enhancing microbial activity, soil structure, and water retention.

• Man-made Fertilizers:

• Manufactured synthetic or chemical fertilizers provide plants a concentrated and faster increase of nutrients.

• Because they are easily soluble and give fine control over nutrient ratios, plants can access them with ease.

• Synthetic fertilizers should not be overapplied since this can result in nutritional imbalances, degraded soil, and contamination of the environment.

4. Application Advice:

• Frequency: Regular, mild fertilizing works better for pineapples than heavy applications. During the growth season, fertilizer should be applied every 6 to 8 weeks.

• Placement: To prevent burning, evenly distribute fertilizers around the plant, avoiding direct contact with the stem.

• Watering: To ensure that nutrients reach the root zone of the plants, water them well after fertilization.

In conclusion, careful consideration of fertilization and nutrition is necessary for pineapple farming to be successful. Growers may optimize fruit output by promoting healthy plant development, weighing the advantages and disadvantages of organic vs synthetic choices, and knowing the particular nutrient requirements of pineapples. Fertilizer dosage modifications will be guided by routine plant health and soil condition monitoring, assuring the long-term viability of pineapple farming.

CHAPTER 7

Techniques For Irrigation

Irrigation techniques must be carefully considered while growing pineapples in order to guarantee ideal growth and fruit development. Maintaining the ideal soil moisture levels and assisting pineapple plants throughout their life cycle depend heavily on proper watering. We'll go into great detail on three important ideas about irrigation methods for pineapple cultivation here: Rainwater Harvesting for Sustainable Farming, Drip Irrigation Systems, and

Watering Quantity And Frequency:

1. How to Create a Watering Schedule:

• Pineapples need regular watering but also like well-drained soil. Create a watering regimen that takes into account the local environment and the particular requirements of the pineapple kind.

• Water the plants often during the growth season to maintain a constantly wet soil. When there is a drought or times of heavy rain, modify the frequency.

2. Preventing Waterlogging

Pineapple plants are prone to developing root rot when placed in damp environments. Finding the right balance between supplying enough moisture and avoiding water stagnation is essential.

• Prevent overwatering and enrich the soil with organic matter to ensure proper drainage.

Systems Of Drip Irrigation:

1. Effective Water Distribution:

• One great way to provide pineapple plants a regulated and effective water supply is by drip irrigation.

• By delivering water straight to the root zone, drip systems minimize water waste and lower the danger of fungal infections linked to overhead irrigation.

2. Precision and Automation:

• Water may be delivered by automated drip irrigation systems at predetermined intervals, guaranteeing a constant moisture content.

• By directing water directly to the root zone, where it is most required, drip irrigation's accuracy contributes to water conservation.

3. Application of Fertilizer:

It is also possible to apply fertilizer precisely by using drip systems. This guarantees that nutrients reach the plants directly, encouraging fruit formation and healthy growth.

Rainwater Collection For Eco-Friendly Agriculture:

1. Rainwater Harvesting for Irrigation:

Gathering and preserving rainwater for use in irrigation at a later time is known as rainwater harvesting.

• Farmers may harness natural precipitation by installing rain barrels or other collecting devices, which lessens their reliance on outside water sources.

2. Sustainability of the Environment:

Rainwater collection lessens the need for groundwater and municipal water sources, which supports sustainable farming methods.

• It also lessens nutrient runoff and soil erosion that are brought on by traditional irrigation techniques.

3. Distribution and Storage:

• Sufficient storage spaces are necessary in order to collect rainwater. Install suitable reservoirs or storage tanks to hold the rainwater that has been gathered.

• Create a distribution system that integrates with the current irrigation infrastructure to effectively feed captured rainfall to the pineapple plants.

To summarize, the best irrigation techniques for pineapple cultivation include adjusting watering schedules to the plant's requirements, setting up effective drip irrigation systems, and taking into account environmentally friendly methods like collecting rainwater. Growers of pineapples may boost yields, encourage healthy plant development, and support environmental conservation by using these techniques.

CHAPTER 8

Controlling Insects And Diseases In Pineapple Production

Typical Pests That Damage Pineapples:

1. Mealybugs:

• Identification: Tiny, white insects on leaves that resemble cotton.

• Damage: Suck sap from plant tissues, which results in growth retardation and discoloration.

• Control: Use insecticidal soaps or introduce natural predators, such as ladybugs.

2. Thrips

• Identification: Wings with fringes on tiny, thin insects.

• Damage: Feed on plant sap, which results in deformed leaves and lower-quality fruit.

• Control: Apply insecticidal sprays or use reflecting mulch to discourage thrips.

3. The nematodes

• Identification: Soil contains microscopic worms.

• Damage: Attack roots, resulting in deficiency of nutrients and withering.

• Control: Apply soil solarization, rotate crops, and use resistant pineapple types.

4. Insects with scales:

• Identification: Tiny, oval-shaped, flat insects on fruit or foliage.

• Injury: Plant sap is drained, weakening the plant.

• Control: Use insecticidal soap or horticultural oil, and prune the afflicted areas.

5. Fruit Flies:

• Identification: Flying, tiny insects that encircle fruit.

• Damage: Fruit contaminated with eggs may ripen and rot too soon.

• Control: Set up traps, maintain proper hygiene, and use pesticides as required.

The Use Of Integrated Pest Management (IPM) And Preventative Measures:

1. Customs and Traditions:

• Select pineapple cultivars resistant to illness.

• To ensure sufficient air circulation, keep plants spaced appropriately apart.

• Make sure the soil drains properly to avoid standing water.

2. Biological Management:

• Introduce beneficial insects such as ladybugs and predatory mites, which are examples of natural predators.

• Use pineapple varieties resistant to nematodes.

3. Catch Crops:

• To entice pests away from the major pineapple crops, use trap crops.

• Regularly check trap crops to determine the number of pests present.

4. Rotating crops:

Rotate pineapple crops in relation to non-host plants to sabotage the life cycles of pests.

• Aids in lowering the accumulation of germs carried by soil.

5. Hygiene:

• Immediately remove and destroy any contaminated plant material.

• Remove weeds from the planting area as they can serve as a haven for pests.

Methods For Identifying And Controlling Diseases:

1. Fusarium Infection:

• Identifying feature: wilting and yellowing of leaves, beginning at the base.

• Control: Rotate your crops, use soil fumigation, and use resistant cultivars.

2. Pineapple Black Rot:

• Fruit with dark, sunken lesions helps identify it.

• Control: Use fungicides, maintain appropriate spacing, and remove infected fruits.

3. Anthracnose:

• Identification: Pink spore masses with small, sunken lesions on fruit.

• Control: Use fungicides, prune appropriately, and keep adequate airflow.

4. Rotting Heart Bacteria:

• Identifying feature: Water-soaked lesions near leaf bases.

• Control: Get rid of diseased plants, maintain good hygiene, stay out of standing water.

5. Virus for Pineapple Mosaic:

• Identification: Leaves with mottled yellow-green patterns.

• Management: Take care of aphid vectors and plant material free of viruses.

A complete strategy to managing pests and diseases in pineapple agriculture may be achieved by combining these measures, which will guarantee healthier crops and higher yields. Efficient management strategies need prompt action and consistent monitoring.

CHAPTER 9

Techniques For Controlling Weeds In Pineapple Growing

The development and yield of pineapples can be greatly impacted by weeds. When it comes to vital resources like water, nutrients, and sunlight, they face competition from pineapple plants. In addition, illnesses and pests that harm pineapples might find a home among weeds. Furthermore, the existence of weeds can impede cultural activities such as cultivation and harvesting, which is why it's critical to put in place efficient weed control methods.

Methods Of Manual Weed Control:

1. Manual Weeding:

• Hand weeding is the process of manually pulling weeds with your hands or with basic equipment. In situations when using equipment is not practicable, it works well for small-scale processes.

2. Grazing and Tilling:

• Frequent cultivation and hoeing aid in breaking up the soil's surface, which inhibits weed growth and keeps persistent weeds from taking hold.

3. Mechanical Weed Control:

• In larger pineapple plantations, mechanical weed control techniques like rotary hoes or tractor-mounted cultivators can be employed to lessen weed competition.

Methods Of Chemical Weed Control:

1. Incipient Herbicides:

• By applying these herbicides to the soil prior to the germination of weed seeds, a barrier that prevents weed development is created. Selecting pesticides that won't damage pineapple plants requires caution.

2. Following Emergence Herbicides:

Post-emergent herbicides (used after weeds have emerged) are designed to kill weeds that are actively developing. It is advisable to use selective herbicides that kill only certain types of weeds without harming pineapple plants.

3. Rotation of Herbicides:

• It's critical to alternate between several herbicides with unique modes of action to reduce the emergence of herbicide resistance in weeds.

Mulching Methods For Weed Control:

1. Natural Mulches:

• By blocking sunlight and inhibiting the germination of weed seeds, organic mulches such as wood chips, straw, or hay can be used around pineapple plants to help limit weed development.

2. Mulches made of plastic:

• You may use plastic mulches to provide a physical barrier that keeps weeds from growing. They also aid in controlling temperature and preserving soil moisture.

3. Living Composts:

• Creating living mulches with cover crops or ground coverings can help reduce weeds and enhance soil health. These live mulches take nutrients away from weeds.

Weed Management That Is Integrated (IWM):

1. Rotating crops:

• Weed life cycles are upset when pineapple is rotated with other crops, which lowers the frequency of particular weed species that are suited to pineapple farming.

2. Biological Management:

• One ecologically benign way to control weeds is to introduce natural enemies of the plant, including herbivorous insects or bioherbicides.

3. Observation and Prompt Identification:

• Weed populations may be regularly monitored to enable early discovery and action, therefore averting extensive infestations.

In conclusion, an integrated strategy incorporating manual, chemical, and mulching approaches, coupled with strategic management strategies, is necessary for efficient weed control in pineapple agriculture. A well-designed weed management approach not only supports healthy pineapple development but also assures sustainable and ecologically responsible farming operations.

CHAPTER 10

Pruning And Thinning Techniques In Pineapple Farming

Pineapple growing requires several cultivation strategies to promote maximum growth and fruit yield. Among these measures, pruning and thinning play key roles in boosting both the quality and quantity of pineapples. Let's explore thoroughly into these topics.

Importance Of Pruning In Pineapple Farming:

1. Enhanced Fruit Quality:

• Pruning is crucial for boosting the overall quality of pineapples. By pruning extra branches and leaves, the plant's energy is diverted towards the growth of fewer but larger and tastier fruits.

• Pruned pineapples tend to yield more consistent and visually beautiful fruits, making them more marketable.

2. Prevention of Diseases:

• Pruning helps generate a more open canopy, promoting air circulation and lowering humidity around the pineapple plant. This, in turn, minimizes the danger of fungal diseases and pests, contributing to a better crop.

3. Resource Allocation:

• Resources like water and nutrients are efficiently divided among the surviving shoots and fruits by smart trimming. Better growth and development of the pineapples are encouraged by this.

4. Aids in Harvesting

Pruning makes it simpler to access ripe pineapples during harvest and removes undesired foliage. This lessens the possibility of fruit damage and streamlines the harvesting procedure.

5. Longevity and Regeneration:

• Regular trimming promotes the longevity of the pineapple plant by encouraging the growth of new branches. This guarantees a continuous cycle of production spanning several seasons.

Fruit Thinning To Enhance Quality:

1. Ideal Distance:

• Thinning is taking off too many fruits so that the remaining ones are spaced properly. This keeps things from getting too crowded and guarantees that every pineapple gets enough sunshine and nutrients to grow to their full potential.

2. Equal Ripening and Size:

• Fruit ripening and size uniformity are facilitated by thinning. For commercial pineapple growing, this is essential since it yields a more uniform product that satisfies consumer demands.

3. Diminished Rivalry:

• The rivalry amongst the remaining fruits for resources like water, nutrients, and sunshine is reduced by thinning out the extra fruits. Larger and more tasty pineapples result from this.

4. Reducing Stress:

• By reducing the amount of stress on the plant, thinning helps it avoid going overboard with fruit production. Pineapples become healthier and more robust as a consequence.

Timing And Methods Of Pruning:
1. Initial Pruning:

The first trimming is usually carried out soon after planting in order to promote the growth of a robust center stalk. This establishes the framework for a pineapple plant with proper structure.

2. Pruning for maintenance:

• Throughout the growth season, routine maintenance trimming is done. This entails trimming off any undesirable branches that can take energy away from the development of fruit, as well as any dead or diseased leaves.

3. After-Harvest Reduction:

• Pruning is done to get rid of any leftover dead leaves and the wasted fruiting stem after harvesting. The plant is now ready for the subsequent growth cycle.

4. Instruments and Methods:

• To prevent further harm to the plant, pruning is best done using clean, sharp instruments. Some techniques are to cut just above nodes or to use the "eye" approach, which involves twisting off undesired shoots by hand.

In summary, thinning and pruning are essential techniques in pineapple production that greatly affect the yield's quality and quantity. A thorough knowledge of the life cycle of pineapple plants and an awareness of the harmony between vegetative growth and fruit production are necessary for the proper use of these approaches.

CHAPTER 11

Harvesting And Handling After Harvest

It may be a fulfilling experience to grow pineapples, and the finest fruit quality is dependent on careful post-harvest management and harvesting techniques. Now let's examine the ideas behind harvesting and post-harvest care in more detail:

Pineapple Maturity Indicators:

1. Shades:

• As they mature, pineapples change color. A bright, golden yellow hue is what you want to look for as it denotes ripeness.

• Steer clear of picking green pineapples as they could not mature correctly and get the sweetness you want.

2. Aroma:

• The base of a ripe pineapple releases a fragrant, delicious scent. The fruit is more developed the greater its scent.

3. Test for Pulling Leaves:

• At the top, gently pull on the inner leaves. It's probable that the pineapple is ripe if they remove readily.

• Take care not to harm the crown since it is necessary for propagation or replanting.

4. Consistency:

• Check for consistent ripeness of the whole fruit. Variations in hue might be a sign of unequal development.

Methods Of Harvesting To Get The Highest Yield:

1. When:

• When pineapples are totally ripe but not overripe, harvest them. Proper timing is essential for the best flavor and texture.

• Five to six months after planting, pineapples are usually ready for harvesting.

2. Utilizing Tools:

• A machete or sharp knife is typically used for harvesting. Carefully cut the fruit off of the stem, leaving a tiny bit of stem intact.

3. Prevent Bruising:

• Take care while handling the harvested pineapples to avoid bruising, which can cause them to deteriorate quickly.

4. Harvesting with Selection:

• Selectively harvest pineapples, beginning with the fully ripe ones. Smaller ones should be let to continue growing and maturing for later harvests.

Tips For Handling And Storing After Harvest:

1. Cleaning:

• Gently wash the gathered pineapples to get rid of any residue or debris. Let them air dry before handling them any further.

2. Humidity and Temperature:

Pineapples should be stored between 45 and 50°F (7 and 10°C) to prevent ripening too quickly.

• To avoid dehydration, keep the relative humidity between 85 and 90 percent.

3. Airflow:

• Encourage enough airflow to lower the possibility of fungus development. Pineapples shouldn't be kept in plastic bags.

4. Divorce:

• Because pineapples release ethylene gas, which can hasten ripening, store them apart from other fruits.

5. Packaging:

• Use the proper packing when shipping or transporting to reduce physical damage and preserve freshness.

6. Verifying the Quality:

• Check pineapples that have been stored on a regular basis for illness, rotting, or ripening. In order to stop problems from spreading, remove any contaminated fruits right away.

These post-harvest handling and harvesting techniques can let you enjoy premium pineapples with the best possible flavor and texture. Whether the fruits are being used for commercial or personal use, proper storage guarantees that they will stay fresh for a longer amount of time.

Summary

To sum up, growing pineapples is a fulfilling endeavor with a distinct set of chances and problems. As we come to a close, it's critical to emphasize important ideas, talk about difficulties, and offer support to aspiring pineapple farmers.

Summary Of The Main Ideas:

1. Pineapple kinds: Successful growing of pineapple requires an understanding of its various kinds. Popular selections include Smooth Cayenne, Queen Victoria, and MD-2; each has a distinct flavor profile and may be adapted to a particular environment.

2. Pineapples grow best on sandy, well-drained soils that have a high level of organic matter. For best growth, they need a tropical to subtropical

environment with warm temperatures. Sufficient sunshine is also necessary for the growth of fruit.

3. Planting and Propagation: Two essential components of pineapple farming are using the right planting techniques and propagation strategies, including using crown cuttings. Successful cultivation depends on using appropriate weed control techniques and maintaining a proper planting spacing.

4. Watering and Irrigation: Pineapples need regular hydration, particularly in arid climates. Effective irrigation techniques, like drip irrigation, can assist preserve water while giving the plants the hydration they require.

5. Management of Nutrients: Fertilization is essential for the growth and development of pineapple fruit. For the best yield, it is crucial to comprehend the nutrient requirements at various

stages of growth and to adopt a balanced fertilization strategy.

6. Management of Pests and illnesses: Pineapple crops are vulnerable to several pests and illnesses. In order to reduce the negative effects of pests and diseases on the crop, it is essential to implement integrated pest management (IPM) tactics, which include biological control techniques and close observation.

7.Harvesting and Post-Harvest management: Keeping fruit quality is influenced by when to harvest pineapples and by using suitable post-harvest management techniques, such as treating with care and choosing the optimal storage conditions.

Obstacles And Remedies In The Pineapple Industry:

1. Pest and Disease Challenges: Mealybugs and fusarium wilt are two pests and diseases that pose

a hazard to pineapple crops. These difficulties can be lessened by routine reconnaissance, early diagnosis, and the deployment of chemical or organic remedies as needed.

2. Market Variations: Growers of pineapples may experience changes in the market that impact demand and prices. To mitigate market volatility, consider diversifying your product offers, looking at value-added choices, and building effective marketing channels.

3. Impact of Climate Change: Pineapple agriculture may be impacted by shifting climatic trends. Investing in climate-resilient cultivars, implementing water conservation measures, and using sustainable agricultural techniques are essential for coping with the effects of climate change.

Motivation For Upcoming Pineapple Growers:

Growing pineapples is a highly promising endeavor for those who are enthusiastic about farming. Potential farmers of pineapples ought to:

1. Accept Sustainability: Using sustainable agricultural methods guarantees the long-term sustainability of pineapple production while also protecting the environment.

2. Constant Learning: To improve knowledge and abilities, stay up to date on the most recent developments in pineapple growing, go to workshops, and interact with agricultural extension agencies.

3. Collaboration and networking: Make connections with other pineapple producers, industry professionals, and specialists in agriculture. Working together can result in the sharing of

resources, expertise, and solutions to problems that arise frequently.

4. Industry Research: To comprehend customer preferences and industry trends, carry out in-depth market research. This information can help direct decisions about manufacturing and support an effective company plan.

5. Adaptability and Resilience: Like any agricultural endeavor, pineapple cultivation is not without its challenges. Having the ability to bounce back and adjust to shifting conditions is crucial for sustained success.

To sum up, growing pineapples is an exciting and rewarding business. Future pineapple producers may support the expansion of this thriving business by putting best practices into practice, overcoming obstacles, and remaining dedicated to sustainability. Remember that perseverance, a strong work ethic, and a love of farming are essential for a successful and fulfilling career in pineapple farming.

THE END